The Low-cost Wooden Duplicator

How to make it; how to use it

by David Elcock

Practical ACTION PUBLISHING

Intermediate Technology Publications 1984

Practical Action Publishing Ltd
27a Albert Street, Rugby, CV21 2SG, Warwickshire, UK
www.practicalactionpublishing.org

© Intermediate Technology Publications 1984

First published 1984\Digitised 2016

ISBN 10: 0 94668 841 9
ISBN 13: 9780946688418
ISBN Library Ebook: 9781780442679
Book DOI: http://dx.doi.org/10.3362/9781780442679

Since 1974, Practical Action Publishing has published and disseminated books and
information in support of international development work throughout the world.
Practical Action Publishing is a trading name of Practical Action Publishing Ltd
(Company Reg. No. 1159018), the wholly owned publishing company of Practical Action.
Practical Action Publishing trades only in support of its parent charity objectives and any
profits are covenanted back to Practical Action (Charity Reg. No. 247257, Group VAT
Registration No. 880 9924 76).

1

INTRODUCTION

The low cost wooden stencil
duplicator is a printing machine
which can be made for use in schools,
colleges and small organizations.
From one inking you can, with
practice, produce over 200 copies of
good quality print. The quality of
the print is nearly as good as that
from much more expensive machines.
The duplicator is made mostly from
wood and you need only simple
woodwork tools to build it. However,
some knowledge of basic carpentry is
necessary to follow the instructions.
It is very easy to use and maintain.
Broken or damaged parts are simple to
replace. The machine is small, made
from wood and is very easy to carry
around.

ACKNOWLEDGEMENTS

The original idea for a low cost,
alternative technology duplicator
came from Dr Salahuddin, of Lipikar
Industries in Bangladesh. The
Industrial Services section of ITDG
has also instigated further
developments with the help of David
Elcock, Senior Lecturer in the
Combined Engineering Department at
Coventry (Lanchester) Polytechnic.
These improvements have been
incorporated in this manual. The
content was designed by Linda
Robinson, Anthea Logan, Magda Lohman
and Sue Hyne, students in the
Department of Graphic Design at
Coventry Lanchester Polytechnic,
under the guidance of Ian McLaren,
Principal Lecturer at the
Polytechnic.

The illustrations in this manual were
drawn by Mirjam Southwell.

Before starting to build the
duplicator please read the following
notes:

1. This instruction manual should be
   read fully before the duplicator
   is built.

   You will notice that specific
   quantities of materials have not
   been stated.

   You can use the templates as
   guides to the basic size, but the

   *You should check the sizes of
   the locally available stencils
   before making the stencil drum.

   width of the stencil drum will
   depend on the width of the
   stencils you will be using.*

2. It should be noted that the
   illustrations are not an accurate
   guide to the sizes of the
   finished parts.

3. It is suggested that hardwood be
   used wherever mentioned to avoid
   parts of the duplicator warping
   in damp conditions. Using
   seasoned wood where possible is
   also advisable.

GENERAL DESCRIPTION

- A single drum is covered with an ink absorbent material over which the stencil is placed (item 1 on Parts and Materials list)

- The pressure roller (item 8) is turned by a handle, this rolls the paper over the stencil and ink is squeezed on to the paper.

- The pressure roller and stencil drum are squeezed together by rubber bands. These can be made by cutting bands from a bicycle tyre inner tube. Extra bands may be required to give greater pressure.

- The hinged covers (items 11, 12 and 13) which give protection for carrying and storage are also the serving and collecting trays for the paper.

- The deflector plate (item 4) guides the paper to the front of the machine.

    When closed, the unit stores the handles, clamp, inking roller and duplicating ink.

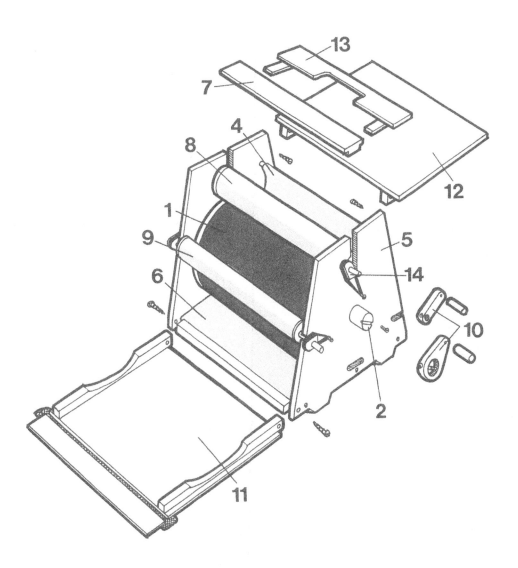

With a little innovation the measurements and materials may be changed to suit locally available materials.

Parts and Materials List

Item No	Description	Material
1	Stencil drum	¾" thick wood 1mm thick plywood or tin plate; felt or other absorbent material, 1" diameter washers. It is important to note that the width of the stencil drum will depend upon the width of the stencils you will use on the duplicator
2	Shaft	1" diameter dowel
3	Fabric grip (not visible on drawing: see page 7)	⁷⁄₁₆" square hardwood flat headed screws scraps of tin plate
4	Deflector plate	Tin plate ⅜" diameter dowel
5	Side panels	¾" thick planed wood round head woodscrews
6	Base	¾" thick planed wood flat headed woodscrews
7	Case handle	¾" thick planed wood
8	Pressure roller	1½" diameter hardwood
9	Ink roller	1" diameter dowel
10	Pressure roller and stencil drum handles	Various pieces of wood
11	Front cover panel	(
12	Back cover panel	((3mm thick plywood (tacks
13	Top cover panel	(
14	Bearings	Hardwood

NB: It is suggested that the wood should be seasoned or given some kind of protective varnish to protect it.

4

HOW TO BUILD IT

Equipment

These are the tools needed to make the
stencil duplicator.

1   A hand drill

2   A wood chisel

3   A plane to smooth the wood

4   A hammer and a screw driver

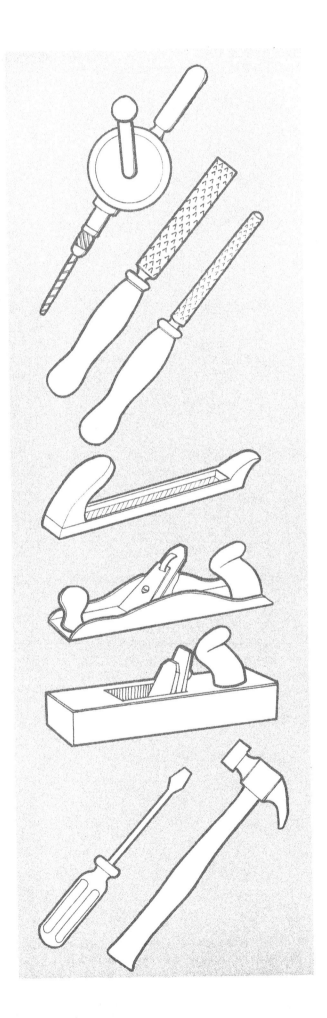

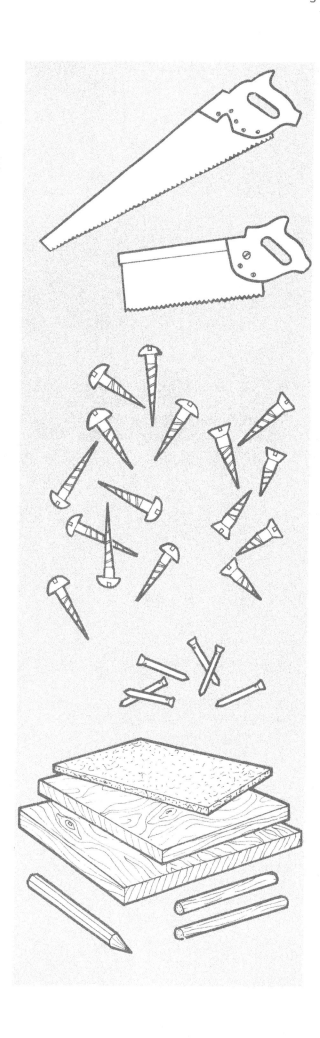

5   A hand saw

6   Several round headed and
     countersunk headed screws and a
     selection of nails

7   A pencil to use with the
     templates and various pieces of
     wood

How to use the templates
(pull out middle pages)

1   Plane both sides of the wood
     smooth

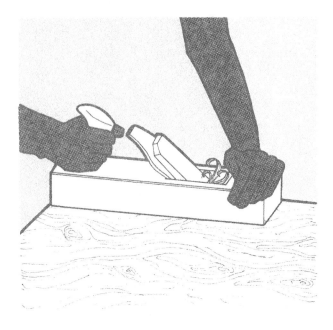

2   Cut out the template and place it
     on the wood. Draw around the
     template and mark on the wood the
     various screw positions and holes

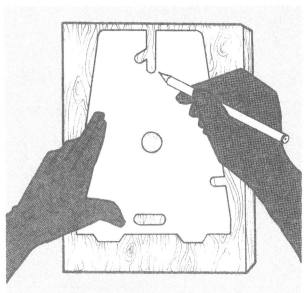

3   Cut out along the lines and sand
     all surfaces before assembling
     the duplicator

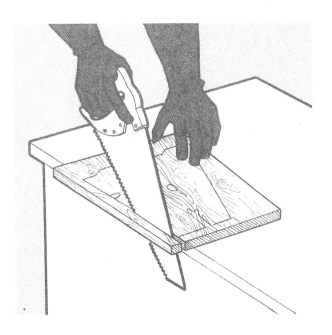

Making the stencil drum

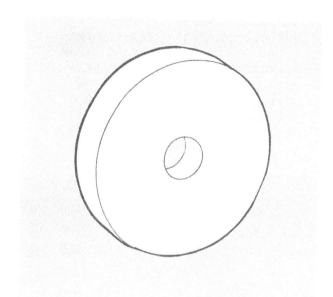

1   Using template 1 cut out together
    three wooden discs (diameter of
    6" and central hole 1" diameter)

2   Cut from each disc a channel for
    putting in the fabric grip (item
    3), ½" deep and 1" wide, drill a
    1" diameter hole in the middle to
    take the shaft

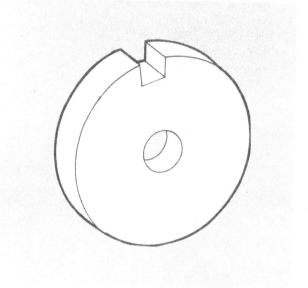

3   The shaft is made from 1"
    diameter dowel cut to a length of
    15". To locate the handle cut a
    groove into each end of the shaft

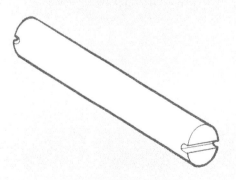

4   One disc should be glued
centrally on the shaft and the
other two discs should be glued
approximately 2¼" from the ends of
the shaft such that their other
surfaces are 10½" apart. This
should leave approximately 2¼" of
shaft at each end

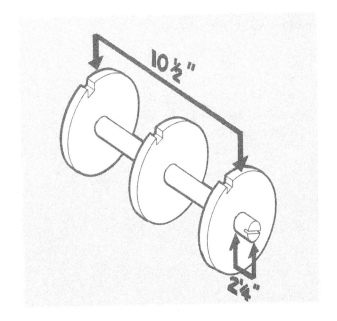

5   Make a U shaped piece of
hardwood, like that shown in the
drawing, 10½" long and ¼" thick.
This should fit in the wooden
discs as shown. Join to the
discs with screws

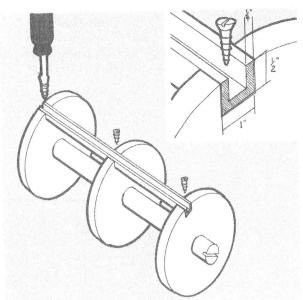

6   Use a piece of tin plate or 1mm
plywood to cover the wooden
discs. Join the tin plate to the
wooden channel section with
tacks. Leave a gap to put in the
fabric grip

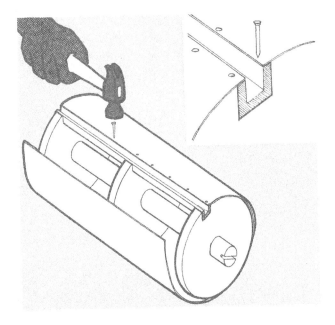

7   Cover the stencil drum with felt
    or absorbent material, allowing
    the material to overlap. Cut out
    the fabric grip from a strip of
    hardwood $\frac{7}{16}$" square and 10½"
    long. Put the fabric grip into
    the channel on the stencil drum.
    This will tighten the fabric
    around the drum. Trim as
    required then secure the fabric
    grip to the stencil drum with
    screws

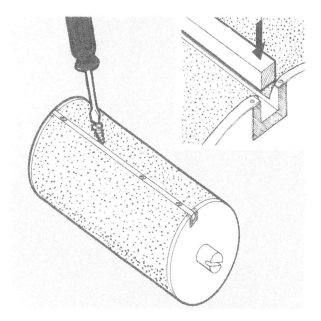

8   Attach three metal grips to hold
    the stencil. These grips should
    be positioned in such a way that
    the head of the stencil you use
    will be held in position
    (different stencils have
    different head fittings).

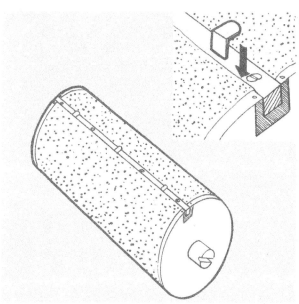

9   Cut out two washers from a
    plastic bottle or something
    similar. Put one on each end of
    the drum shaft to allow the
    stencil drum to turn freely

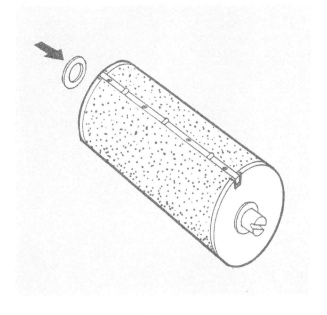

Making the side panels and base

1   Using template 2 cut out together
    two side panels from ¾" thick
    planed wood. Cut out the various
    holes and mark on the positions
    for the screws

2   Using template 3 cut out the base
    from ¾" thick planed wood and
    make a groove ¼" deep for
    inserting the deflector plate
    (item 4)

3   Join one of the side panels to the
    base using three screws. Put the
    stencil drum shaft into one of the
    side panels

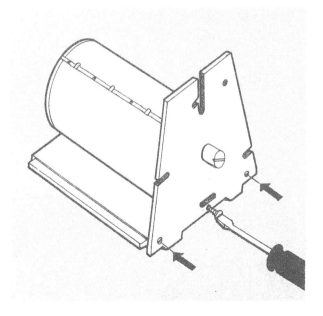

4   Join the second side panel to the
    base using three screws

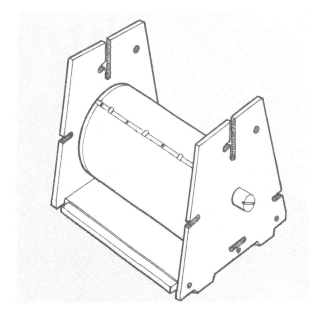

Deflector plate

Cut out a piece of tin plate
approximately 10" wide. Bend the
tin plate around the drum to get
a rough curve. Fit one end of
the tin plate into the slot in
the base and curve the tin plate
around the back of the drum.
Glue a piece of ⅜" diameter
dowel into the holes at the top
of the side panels and bend the
tin plate around the dowel

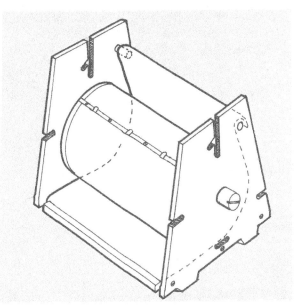

12

Making the pressure and inking rollers

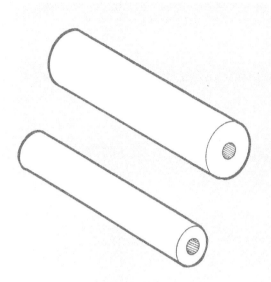

1   Make the pressure roller from a 1½"
    diameter piece of dowel which is 10⅞
    long. Similarly with the inking
    roller, but this time with 1"
    diameter dowel. In each roller
    make a ⅜" diameter hole at each
    end to a depth of 2"

2   Glue the holes and insert shafts
    made from four 2½" lengths of
    dowel

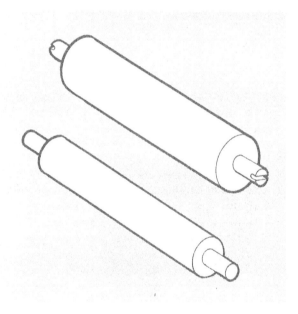

3   Cover both rollers with material,
    preferably felt

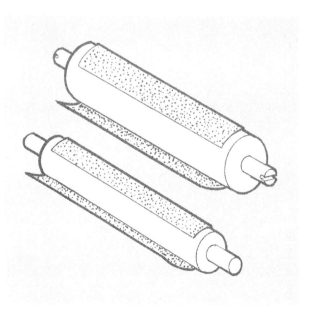

Bearings

Cut out four bearings using
template 4 from hardwood to a
width of approximately ⅝" . Slot
two of the bearings onto the
pressure roller shaft. Put
screws in position shown in the
drawing and then hook rubber
bands around the bearings and
screws. When using the inking
roller attach the bearings as
above

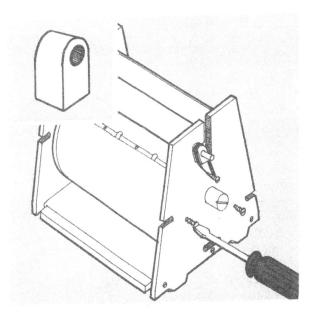

Pressure roller and stencil drum
handles

1   Cut out the handles using the
    templates 5 and 6. Mark on the
    screw position and cut out the
    holes

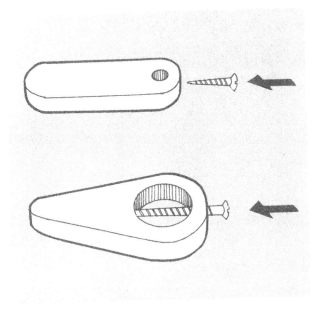

2   To attach the handles to the
    shafts put a screw of minimum
    length 1¾" through the end of the
    handle

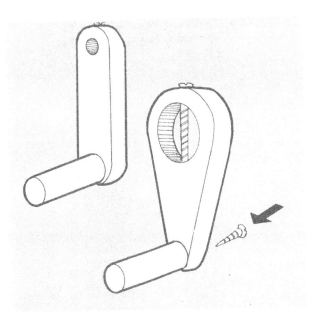

3   Join a length of dowel to the
    handle and attach it using a
    countersunk headed screw

Making the cover panels

(This is an option and is not needed
for the duplicator to work. A brief
description is given below, also
refer to general description drawing
on page 2)

1   Using template 7 cut out the
    cover panels from 5mm plywood.
    Glue and tack the supports to the
    main pieces. For the front cover
    panel (item 11) make a hinge from
    a piece of material or something
    similar and fix it with glue.
    Join the panel to the base using
    two long screws which pass
    through the holes in the cover
    panel supports

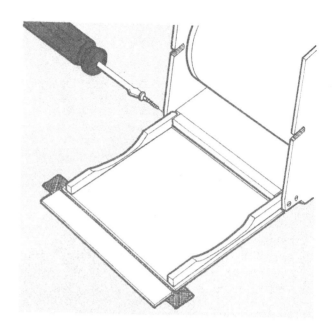

2   Join the back cover panel (item
    12) to the side panels using
    round headed screws. Put the
    screws through the slots to allow
    the panel to move up and down

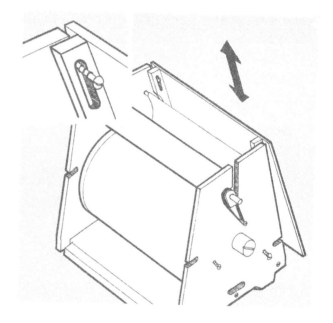

3   The top cover panel (item 13)
    should fit underneath the case
    handle (item 7)

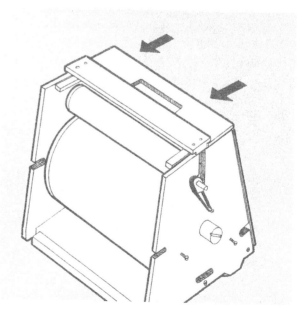

HOW TO USE THE DUPLICATOR

Inking

1   Lift the pressure roller into the
     non-printing position

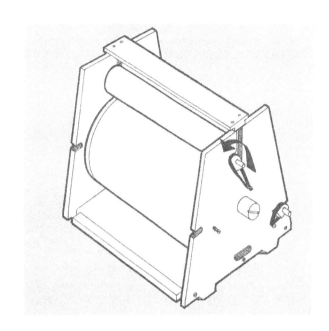

2   Remove the inking roller from its
     store position at the back of the
     duplicator and attach as shown

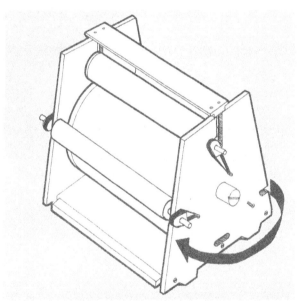

3   Attach the handle to the stencil
     drum

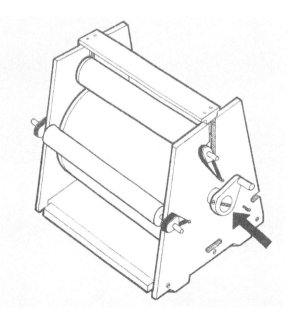

4   Put the ink along the roller in
     the trough between the roller and
     drum

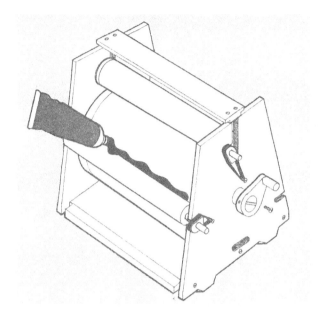

5   Turn the stencil drum handle back
     and forth to spread the ink over
     the roller. Make sure you only
     ink the area which will be
     covered by the stencil (you do
     not even need to ink the outer
     edges of the roller as you will
     not type to the very edge of the
     stencil). Do not use too much
     ink and do not cover the complete
     stencil drum or you will ruin the
     stencil and rollers

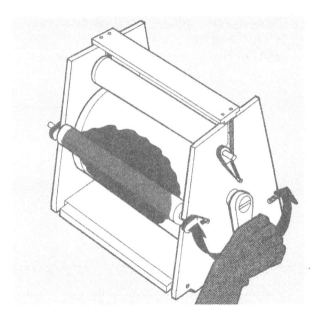

6   Remove the inking roller and
     store it in the back of the
     duplicator

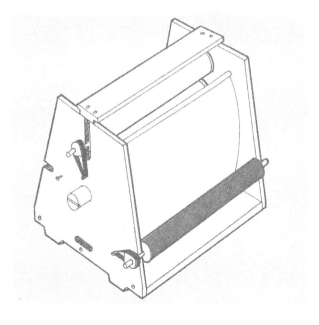

The stencil

7   Remove the carbon from between the
     stencil and the backing paper

8   Hook the stencil with the backing
     paper attached on to the metal
     clips on the stencil drum

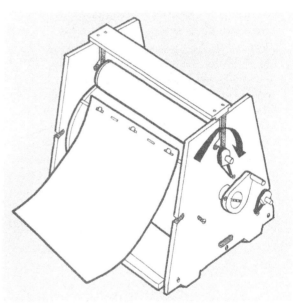

9   Put the pressure roller into the
     working position

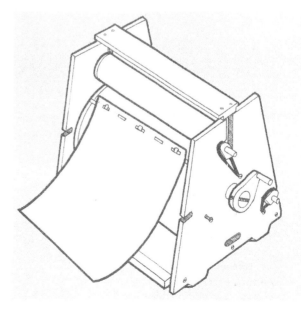

18

10 Remove the handle from the stencil
    drum and attach the pressure roller
    handle to the pressure roller

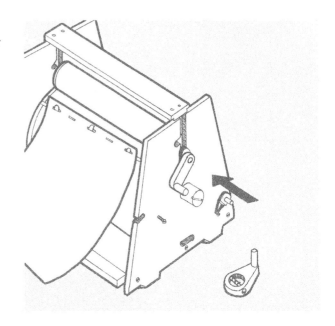

11 To test if the ink is passing
    through the stencil turn the
    pressure roller anti-clockwise.
    If not enough ink is passing
    through the stencil add extra
    rubber bands or rub your fingers
    over the backing paper to squeeze
    the ink through. Once this is
    done the ink will flow
    continuously

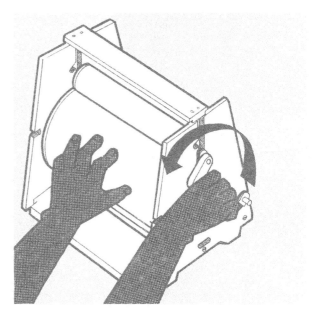

Taking the copies

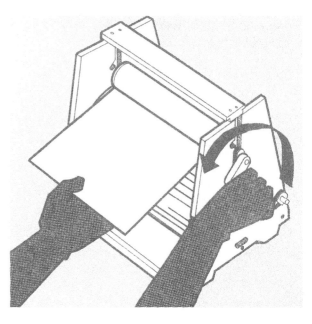

12 Remove the backing sheet and then
    put a piece of blank paper
    between the pressure roller and
    the stencil drum. Turn the
    pressure roller anti-
    clockwise

13 The printed copy will be
    deposited in the tray

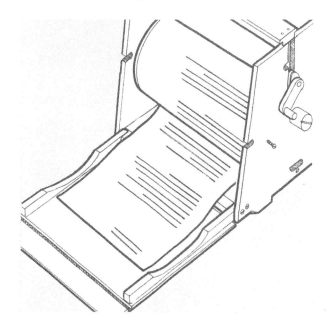

CAUTION

Do not turn the pressure roller
without paper over the stencil because
it will put ink on to the pressure
roller and on to the back of further
copies

If this happens clean the ink off the
pressure roller

After use

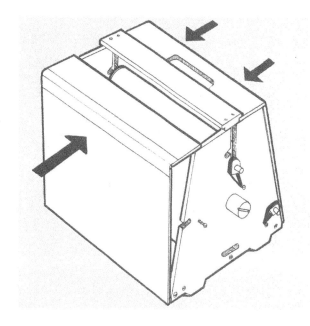

14 Remove the stencil from the
    machine using top edge under
    clips

15 Take the handle off the pressure
    roller and fold the covers down,
    put the handle inside before
    closing

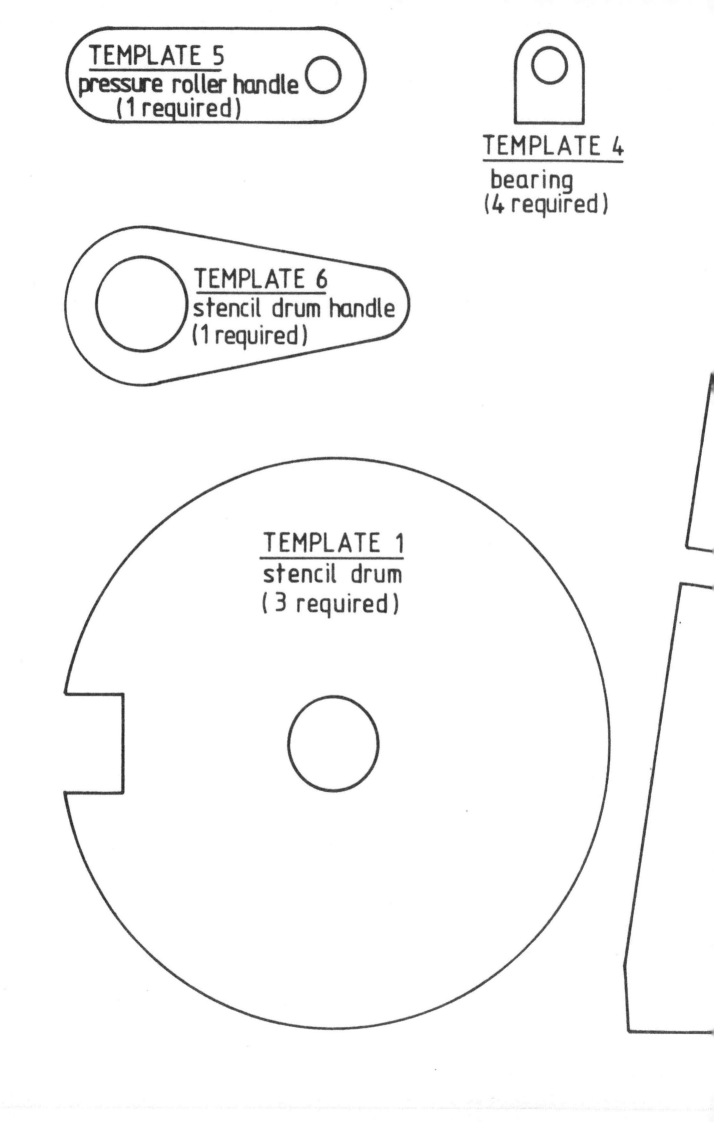

TEMPLATE 5
pressure roller handle
(1 required)

TEMPLATE 4
bearing
(4 required)

TEMPLATE 6
stencil drum handle
(1 required)

TEMPLATE 1
stencil drum
(3 required)

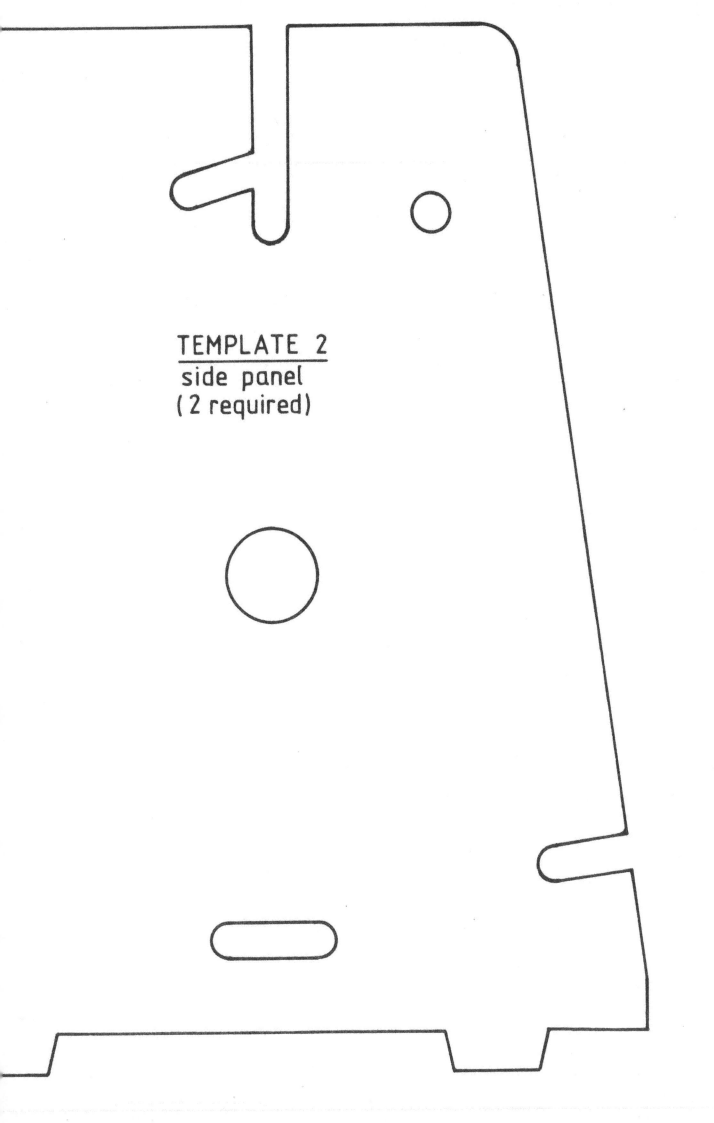

TEMPLATE 2
side panel
(2 required)

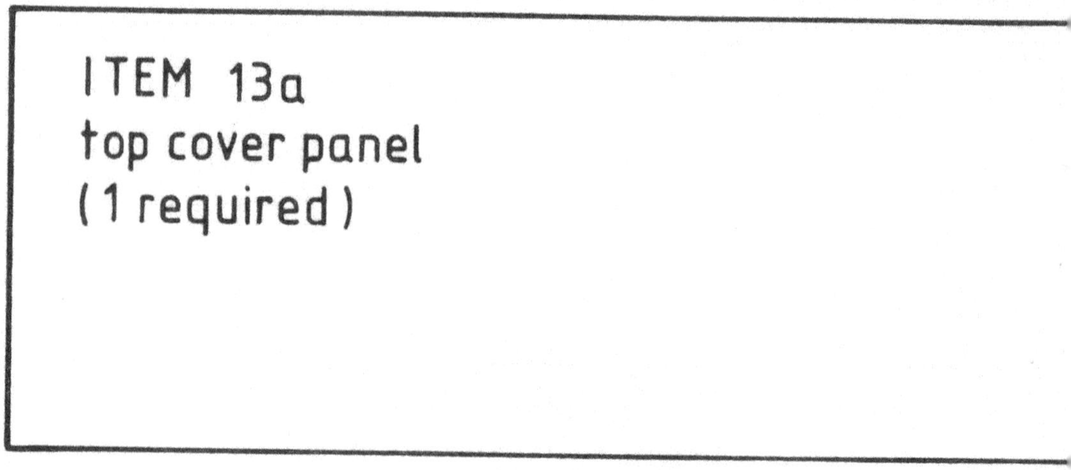

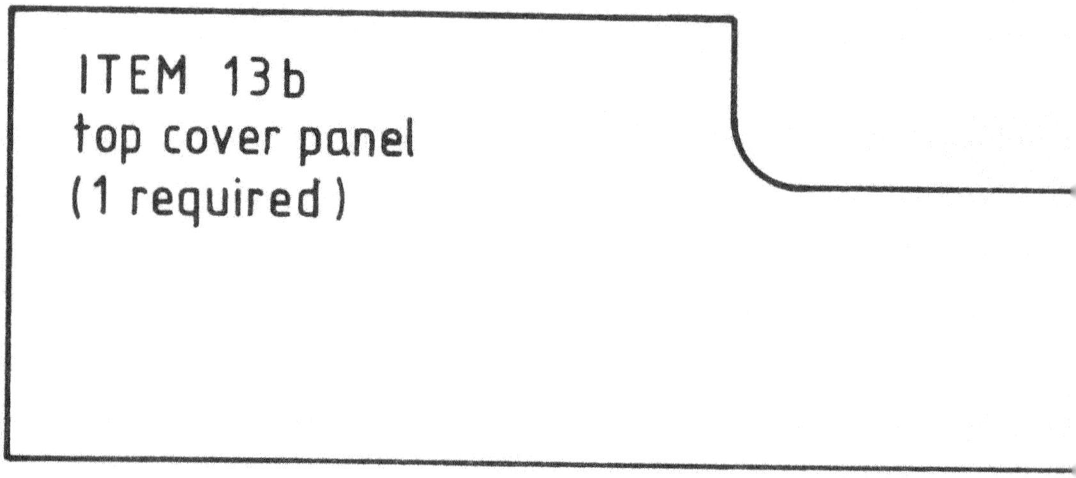

ITEM 13a
top cover panel
(1 required)

ITEM 13b
top cover panel
(1 required)

part of front
(2 required)

TEMPLATE 7 – continued

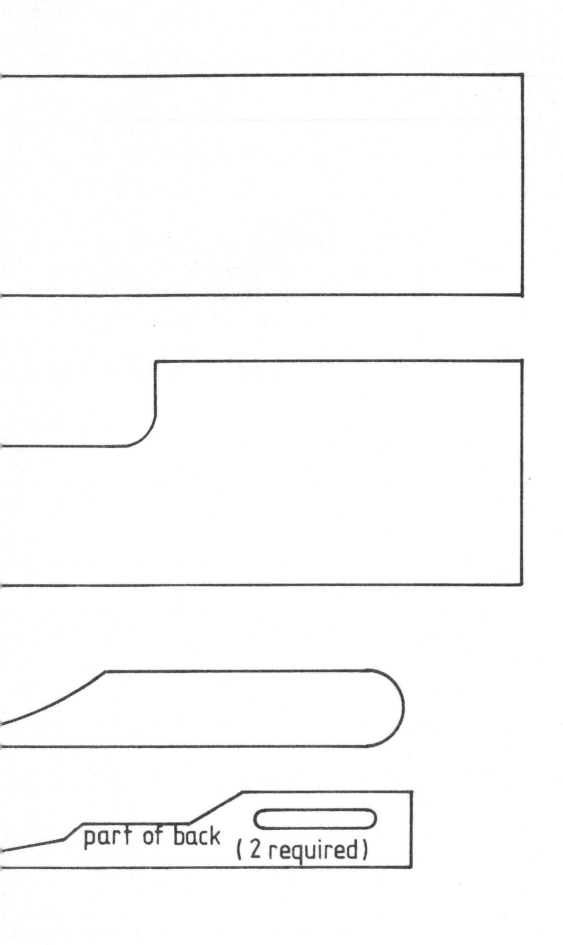

part of back (2 required)

TE
ITEMS 1
(2

cover panel
red)

TEM

bas

ITEM 7
case handle
(1 required)

E 3

equired)

end view
showing material shape

Milton Keynes UK
Ingram Content Group UK Ltd.
UKHW050424080824
1191UKWH00065B/257